Samsung S9 Secrets

Your Guide to Getting the Most Out of Your Samsung S9 and S9 Plus

Rodger Peck

© **Copyright 2018 - All rights reserved.**

The content contained within this book may not be reproduced, duplicated or transmitted without direct written permission from the author or the publisher.

Under no circumstances will any blame or legal responsibility be held against the publisher, or author, for any damages, reparation, or monetary loss due to the information contained within this book. Either directly or indirectly.

<u>Legal Notice:</u>

This book is copyright protected. This book is only for personal use. You cannot amend, distribute, sell, use, quote or paraphrase any part, or the content within this book, without the consent of the author or publisher.

Disclaimer Notice:

Please note the information contained within this document is for educational and entertainment purposes only. All effort has been executed to present accurate, up to date, and reliable, complete information. No warranties of any kind are declared or implied. Readers acknowledge that the author is not engaging in the rendering of legal, financial, medical or professional advice. The content within this book has been derived from various sources. Please consult a licensed professional before attempting any techniques outlined in this book.

By reading this document, the reader agrees that under no circumstances are is the author responsible for any losses, direct or indirect, which are incurred as a result of the use of information contained within this document, including, but not limited to, —errors, omissions, or inaccuracies.

④ shop Bluetooth Headset

① setup Samsung Account Finger Print Scanner

② Land's End - suit

③ Meetup Group of Investors (Fridays)

Table Of Contents

Introduction

Chapter One: Basics of your Phone

  Comparison to Other Devices

  S9 and S9 Plus Specifications

    Dimensions & Weight:

    Display

    Camera

    Rear Camera

    Front & Rear Camera

    Video Recording

  Performance

  Memory

  Network & Connectivity

  Expandable Memory & SIM Card

  OS

  Audio

Audio playback format

~~FIX~~ — **Bluetooth** — *FIX kitchen!*

Recording

Video

TV Connection

Water Resistance

Battery

Charging

Hardware

   Front

   Back

Unpacking

Chapter Two: Setting up your Phone

Assembling

Installing SIM and Memory Card

Charging Battery

Turning the Phone on and Setup Wizard

Locking and Unlocking:

Adding a Google Account:

※ **Transferring Data** — To A chromeBook Look Into

Set Up Voicemail:

Chapter Three: Navigating the Device

Navigation

Tap
Double-Tap
Touch and Hold
Drag

Touch Sensitivity

Navigation Bar

Show and Hide Button
Background Color
Button Layout

Home Screen

App Shortcut

Wallpapers

Themes

Icons

   Icon Frames:

Widgets

Home Screen Settings

Easy Mode

Status Bar Tricks

Notification Panel

Edge Screen

Apps Edge

People Edge

Smart Select

Edge Lighting

Edge Lighting Advanced Settings

Manage Edge Lighting Notifications

Quick Reply

Chapter Four: Apps

Camera and Video

Shooting Mode

    Rear Camera

    Front camera

Top 5 Camera Features

Galaxy Apps

Gallery

Samsung Health

Chapter Five: Special Features

Bixby

Setting Up Bixby

Bixby Vision

Top 5 Hidden Bixby Features

Securing Your Phone

    Face Recognition

    Iris Scanner

    Fingerprint Scanner

    Intelligent Scan

    Multi Window

Samsung Pay

Customizing your Screen

AR Emoji

Customize Messages App

Conclusion

Introduction

Firstly, I want to thank you for choosing this book, Samsung S9 Secrets: Your Guide to Getting the Most Out of Your Samsung S9 and S9 Plus.

Samsung launched the Galaxy S9 and S9+ in March 2018, and since then it has been praised for its great processor, the new Android Oreo system, and especially its excellent camera that comes with dual aperture.

The phone has a great feel to it and comfortably fits in the hand. The best part about the phone is its display, which is 5.8 inches for the S9 and 6.2 inches for the S9 plus. The display covers the entire screen and gives a tremendous feel to the entire phone.

The phone comes with a great octa-core processor, which ensures that the phone never lags. Even when you have multiple apps on at the same time, including some massive games, the phone will continue to function well. The RAM space is also very high for both versions of the phone. There is also enough internal storage available to ensure that you don't need any external help. However, an external microSD card is available that allows you to expand the memory of your phone to 400 GB.

The phone comes with the new Android Oreo 8.0, which puts it at the front of cutting-edge technology. Other than that, Samsung has introduced some great features to make the phone better, such as Bixby, dual aperture, and water and dust resistance.

This book is a guide for using the S9 and S9 plus. In it, you will find everything that you need to know about setting up and navigating through your new device. The book will provide you with not just basic information on using the phone, but also some amazing tips and tricks that will help you to get the best out of your new device.

The S9 and the S9 plus only have a few differences; if you want a phone with a better RAM and more enhanced qualities, then you should go for the S9 plus.

Chapter One:

Basics of your Phone

The Samsung S series has revolutionized the world of smartphones with its brilliant phone aesthetics, camera, and processors. Samsung with its S series has been the flag-bearer of innovation in Android phones and has been competing with its long-term rival Apple for a decade.

With the 9th generation of the S series – the Samsung S9 and S9 plus – what we have is a brilliant phone that seeks to give you anything that you might require in a phone. The build of the phone is extremely aesthetically pleasing, and I am sure your friends will be impressed as soon as they see this phone in your hand.

There are three particular features of the 9th generation S series that make it worth buying. <u>The first</u> is the Dual Aperture camera that turns your phone into the human eye. The camera is so good that it adapts to light like the human eye; the camera adjusts itself based on the amount of light entering the camera, and this ensures that you get good high definition pictures no matter if it's day or night.

over Rated

After this, you have Bixby. Bixby responds to whatever voice commands you give and will help you to set up your Samsung device. With Bixby you don't even have to touch the phone to use it – you can run the whole system using Bixby.

useless
I Hate

— I like google Better

2   SAMSUNG GALAXY TAB 4

The S9 and S9 Plus also ensure that your phone will always be secure and it's great for Samsung since it gives the customer much more flexibility to set guidelines to ensure their privacy. You can set face recognition – with this the front camera of the phone will scan your face to open it. You can also set a fingerprint scanner, not just to unlock your phone, but to open various apps and to even secure your Samsung Account.

All of these features make the Samsung S9 and S9 plus unique, and you'll fall in love with this phone. We'll go into detail about these various features in the coming chapters.

Comparison to Other Devices

Apple iPhone X – The significant difference between these two phones is in the processor. The iPhone X has a Hexacore, while the S9 has an Octacore. This means that the S9 is made up of 9 cores, while the iPhone X is made up of 6. The impact of this can be seen in the performance and longevity of both phones.

Another major difference is that the iPhone X does not have a fingerprint scanner and has a slightly smaller battery compared to the S9.

OnePlus 6 – The OnePlus 6 has a slightly larger screen than the S9, and it also has a somewhat better camera at 16

MP. OnePlus 6 also has a slightly larger battery by 300 mAh. There aren't many differences between the two, other than this, and both of them have the same processors and operating systems.

Note 8 – If you're a Samsung fan you will be confused between these two phones. Note 8 and S9 both have the same design, except the slightly more metallic finish of the S9. Both the phones look amazing and are almost the same size.

S9 is running the Oreo system right out of the box, while the Note 8 will probably get the Oreo update in a short time, so it's not a selling factor for the S9. Note 8 has a lot more disadvantages compared to the S9, especially in two important aspects – hardware and camera.

When it comes to hardware, the Note 8 does not have stereo speakers, and that can be a major discomfort if you want to hear songs out loud. The Note 8 is also more difficult to handle since it's more prominent in style and the fingerprint scanner is a little difficult to reach.

Note 8 has a great camera, but it's nothing compared to the cameras of S9 and S9+. The camera is the selling point of S9, and if you want a phone to take some great pictures for your Instagram or Snapchat, then S9 is the way to go.

S9 and S9 Plus Specifications:

Colors: Midnight Black, Lilac Purple, Coral Blue, Titanium Gray, Burgundy Red and Sunrise Gold

Dimensions & Weight:

Galaxy S9

- Dimensions: 147.7 x 68.7 x 8.5 mm
- Weight: 163 g

Galaxy S9+

- Dimensions: 158.1 x 73.8 x 8.5 mm
- Weight: 189 g

Display

Galaxy S9

- 5.8" Quad HD+ Super AMOLED (2960x1440)
- 570 ppi

Galaxy S9+

- 6.2" Quad HD+ Super AMOLED (2960x1440)
- 529 ppi
- Default resolution is Full HD+ and can be changed to Quad HD+ in Settings.
- Infinity Display: a near bezel-less, full-frontal, edge-to-edge screen.

What is "Bezel"

- Screen measured diagonally as a full rectangle without accounting for the rounded corners.

Camera

- Front Camera: 8MP AF sensor
- Pixel size: 1.22μm
- Sensor size: 1/3.6"
- Sensor ratio: 4:3
- F1.7 aperture
- FOV: 80°
- Selfie focus
- Wide Selfie

Rear Camera ?

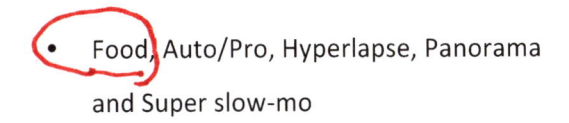

- (Food), Auto/Pro, Hyperlapse, Panorama and Super slow-mo

(Super slow-mo only supports HD resolution. Limited to 20 shots per video with approximately 0.2 seconds of recording and 6 seconds of playback for each shot.)

Galaxy S9

- OIS (Optical Image Stabilization)
- Selective Focus (background blur effect)
- Digital zoom up to 8x

Galaxy S9+

- (Dual camera)   where or How do I Activate? switch Between Both

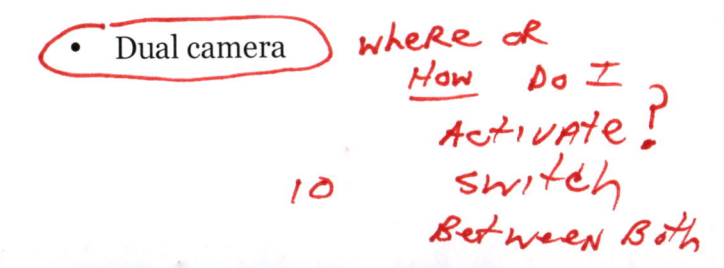

10

- Dual OIS (Optical Image Stabilization)
- Optical zoom at 2x
- Digital zoom up to 10x
- Live focus with bokeh filters (background blur effect)
- Dual Capture

Front & Rear Camera

- VDIS (Video Digital Image Stabilization)
- Grid lines
- AR Emoji
- HDR (High Dynamic Range)
- Location tags

- Motion photo

- Full view

- Timer

- (Stamps) — wHAt? How Do I Activate?

- Filters

- Floating camera button

- Quick launch

(Apps that do not support animated GIFs may still send AR Emoji stickers as a still image.)

Video Recording

- 4K video recording at 30 fps or 60fps

- 1080p HD video recording at 30 fps or 60 fps

- QHD video recording at 30 fps

- 720p HD video recording at 30 fps

- Slow motion video support 1080p at 240 fps

- Super Slow-mo video support 720p at 960 fps

- Hyperlapse video support 1080p

- Digital zoom up to 8x (Galaxy S9) or up to 10x (Galaxy S9+)

- VDIS (Video Digital Image Stabilization)

- Playback zoom

- High CRI LED Flash

- Face detection
- Continuous autofocus video
- Take 9.1-megapixel still photos while recording 4K video
- Tracking AF
- Video location tags

(Super Slow-mo only supports HD resolution. Limited to 20 shots per video with approximately 0.2 seconds of recording and 6 seconds of playback for each shot.)

Performance

AP

- 10nm 64-bit Octa-Core Processor *2.8GHz + 1.7GHz (Maximum Clock Speed, Performance Core + Efficiency Core)

- 10nm 64-bit Octa-Core Processor *2.7GHz + 1.7GHz (Maximum Clock Speed, Performance Core + Efficiency Core)

(May differ by country and carrier.)

Memory

Galaxy S9

- 4GB RAM

- 64GB

✓ Galaxy S9+ mine

- 6GB RAM

- 64GB

(May differ by country and carrier.)

(Actual user memory may be less due to pre-installed operating system/software and may change after software updates.)

Network & Connectivity

- Enhanced 4x4 MIMO/CA, LAA, LTE Cat.18

- Wi-Fi 802.11 a/b/g/n/ac (2.4/5GHz), VHT80 MU-MIMO, 1024QAM

- Bluetooth® v5.0 (LE up to 2Mbps), ANT+, USB type-C, NFC, location (GPS, Glonass, Galileo, BeiDou)

(Galileo and BeiDou coverage may be limited. BeiDou may not be available for certain countries.)

[handwritten notes: issues with being backward compatable. Had to install a Bluetooth app to run Sony stereos...!!! 17]

Expandable Memory & SIM Card

- Single SIM model: one Nano SIM and one MicroSD slot (up to 400GB)

- Dual SIM model (Hybrid SIM slot): one Nano SIM and one Nano SIM or one MicroSD slot (up to 400GB) *Find online for S4 Tablet*

OS

- Android 8.0 (Oreo)

Audio

- Stereo speakers tuned by AKG

- Surround sound with Dolby Atmos technology (Dolby Digital, Dolby Digital Plus included)

- Ultra High-Quality Audio Playback

- UHQ 32-bit & DSD support PCM: Up to 32 bits

- DSD: DSD64/128

(DSD64 and DSD128 playback can be limited depending on the file format.)

Audio playback format

- MP3, M4A, 3GA, AAC, OGG, OGA, WAV, WMA, AMR, AWB, FLAC, MID, MIDI, XMF, MXMF, IMY, RTTTL, RTX, OTA, APE, DSF, DFF

Bluetooth — BLOWS
TRY this tRick AgAIN

- Dual Audio: connect two Bluetooth devices to the Galaxy S9 or S9+ to play audio through the two devices simultaneously. (The two connected devices may exhibit a slight difference in sound output.)

HA! THAT IS AN UNdeRstAtement

- Scalable Codec: Enhanced Bluetooth connection under ambient radio frequency interference. (Available only for certain accessories made by Samsung.)

FIND EAR Piece Bluetooth

Recording

Recording quality is improved with the High AOP Mic that minimizes distortion in noisy environments.

(AOP: Acoustic Overload Point)

- Bundled Earphones
- Pure sound tuned by AKG
- Hybrid canal type

- 2way dynamic unit

Video

Video Playback Formats

- MP4, M4V, 3GP, 3G2, WMV, ASF, AVI, FLV, MKV, WEBM

TV Connection

- Wireless: Smart View (Miracast 1080p at 30 fps, mirroring support available for devices supporting Miracast or Google Cast)

(○) *FIND A CABLE*

- With cable: supports <u>DisplayPort</u> over USB type-C. Supports video out when connecting via HDMI Adapter (DisplayPort 4K 60 fps).

Water Resistance

- IP68

(Based on test conditions of submersion in up to 1.5 meters of fresh water for up to 30 minutes.)

Battery

<u>Galaxy S9</u>

- Battery Capacity - 3000mAh

Battery Life

- MP3 playback (AOD on): up to 48 hrs.

- MP3 playback (AOD off): up to 80 hrs.

- Video playback: up to 16 hrs.

- Talk time: up to 22 hrs.

- Internet use (Wi-Fi): up to 14 hrs.

- Internet use (3G): up to 11 hrs.

- Internet use (4G): up to 12 hrs.

Galaxy S9+

- Battery Capacity: 3500mAh

Battery Life

- MP3 playback (AOD on): up to 54 hrs.
- MP3 playback (AOD off): up to 94 hrs.
- Video playback: up to 18 hrs.
- Talk time: up to 25 hrs.
- Internet use (Wi-Fi): up to 15 hrs.
- Internet use (3G): up to 13 hrs.
- Internet use (4G): up to 15 hrs.

Charging

- Fast charging on wired and wireless

- Wireless charging compatible with WPC and PMA *— Lookup TERMS*

Hardware

In this section, we will learn about the underlying hardware of your phone. You will learn about the different buttons, and other things on your phone and why you need them.

Front

- Proximity Sensors – At the top left, you have the proximity sensors that close the screen when you're on a call and turn the screen back on when the phone is away from your body.

- Power Button – You can use this button to close the screen. Press and hold this button to see the power options. If you press it twice quickly, it will launch the camera.
- Volume Keys – You can use the volume keys to adjust the volume for ringtone, media, and notifications. You can also press this button in the camera app to take a picture.
- Bixby Key – This key is on the left side of the screen and clicking it will launch Bixby.

Back

- On the back, you can see multiple things such as the cameras, headphone jack, flash, fingerprint scanner, speaker, and microphone.

Unpacking

Before we move on to setting up the device, we will take a look at what is in the box and how to unpack everything. It's fairly easy to open the box. Use a knife and cut the seals.

Your product box should contain the following items:

1. Samsung Galaxy S9/S9 Plus

2. USB Charging cable and power adapter
3. Quick Reference Guide
4. Earphones
5. USB Connector (USB Type-C)
6. Eject Pin
7. Micro USB Connector

Chapter Two: Setting up your Phone

Now that you have the Samsung Galaxy S9/S9+ let's look at how to set up your phone. In this section, we will work through a few things – how to assemble your phone, and how to configure the basics of your phone so you can start using it.

Assembling

There are three things that you have to do in the assembly phase. You have to install the SIM card, then the memory card (if you want to), and finally, charge the battery.

Find YouTube videos

Installing SIM and Memory Card

CAN I copy A SIM CARD? *USE ON NOTE 5?*

A SIM Card is a small chip that is provided by the cellular network operator that allows you to contact other people via calls and texts. It's the chip that has your number and other details encrypted on it. Once you put the SIM Card into your phone, you'll be able to contact people with your number. If you don't have a <u>SIM Card,</u> get one from a local shop and if you already have one, make sure that it's the right size.

Done 256 Gig <u>CARD</u>

You can also install an optional microSD ✓memory card, which will give you extra space on your device. Both the S9 and S9+ come with internal memory on which you can store things, but

Auto stamp <u>Photos</u>   *31 problems*

sometimes you might need additional memory, especially if you plan to store a lot of things on your phone. Now, to install both the SIM and the microSD card, look in the box, and you'll find a <u>removal tool</u> that you can use to eject the internal card tray in which the SIM and the microSD card go.

[handwritten: FiNd TooL]

Follow these steps to install the SIM and microSD card:

1. Use the removal tool to take out the internal card tray. Take the removal tool and push it into the hole that is on the internal card tray. You can find this card tray at the top of your phone. Just push the removal tool into the hole until the tray pops out.

[handwritten: How ABout A simple PApeR cLip?]

2. Place the SIM card and the microSD card on the tray. The gold contact of the SIM Card should be facing down. The same applies to the microSD card.

3. Slide the tray back into the slot, and as soon as you hear a click, your work is done.

Charging Battery

Samsung S9/S9+ function by using a rechargeable battery that continues to drain while you're using the phone. You can recharge the battery by using the charging head and USB Type-C cable that is provided in the box.

Your battery must already be half full, but it's still important to charge the phone completely before you start using it. This is only important for the first time that you're using the phone. Also, <u>don't ever use any other charger to charge your phone because it might damage your phone, since the charger might not be built to the specifications of your phone.</u>

What?! Well, how will I know

To charge your phone, follow these steps:

1. Connect the USB Type-C cable to the phone.

2. Insert the cable into the charging head.

3. Connect the charging head to a standard outlet and switch it on. Close the switch and then take

out the cable and the charging head once the phone is completely charged.

It's normal for the charger and the phone to heat up when charging and it does not mean that your phone will get damaged. If your phone gets too hot, stop charging and keep it in a cool place for a few hours.

Turning the Phone on and Setup Wizard

After you are done assembling your phone, the next step is to turn it on and then go through the Setup Wizard so you can configure the basic settings of your phone.

If you want to turn your phone on or off, follow these steps:

- Hold the power button down until the device turns on.

- Hold the power button down for a few seconds, and you'll see a few options on the phone. Click on the Power off command and your phone will shut off.

- When you turn your phone on for the first time, the first thing you will see is the Setup Wizard. The Setup Wizard helps you to configure basic settings so you can start using your phone without any hassle.

You don't need much guidance for the Setup Wizard since there are prompts that you will see on the screen. These prompts will ask you to pick a default language, connect to a Wi-Fi Network, and set up different accounts. You will also get to learn about the different features of your device.

Locking and Unlocking:

Locking and unlocking the phone depends on the security features that you have set up on your phone. By default, you can lock and unlock by doing this:

1. Clicking on the Power Button will lock the phone.

2. Click on the power button (or) double-tap the Home button (the square button at the bottom) and

I AM NOT DOING Right Double TAP!!

37

then swipe your finger across the screen.

Swipe is the default screen lock on your phone, and if you want to change it, we'll discuss it in the next chapter.

Adding a Google Account: ✓ Done

Your phone runs on Android, and you will need to add a Google account to use your phone.

- From the home screen, swipe up and then click on Apps.

- Go to settings – Cloud and Accounts – Accounts.

- Tap Add Account – Google.

- Similarly, you can also add a Samsung Account and an Email Account.

Transferring Data

You might have some critical data on your old phone like contacts, texts, and other things that you want on your new phone. There are two ways to transfer all of the data from your old phone to your new phone.

- You can either use the On-the-Go adapter included in the box.

- Just connect the adapter to both the phones, and then on the old device select the Media Device

(MTP) option when the USB cable prompt comes.

If this isn't suitable for you, you can use the <u>Smart Switch App.</u> If you want to use Smart Switch:

- From the home screen, swipe up and click on Apps. Go to settings – cloud and accounts – Smart Switch.

Find App on Phone

You'll get some prompts on the screens. Just follow them to complete the transfer.

Set Up Voicemail:

- On the home screen, click on the green phone option.

- Now click on the voicemail button.

- A tutorial will begin that will help you to record a small greeting, set up a password, and also record your name.

You have finished assembling your phone and configuring the basics. It's time to navigate the phone and learn about what your phone can do.

Chapter Three: Navigating the Device

Navigating the device is all about learning what your device can do and how you can use the different features. In this section, I will teach you how to use your device, its various features, and some neat tricks and tips that will help you to get the most out of your phone.

Navigation

To navigate anywhere on your phone all you need to do is lightly tap the screen. It's important to remember that the screen is delicate and if you touch it with a lot of force or with a sharp object, it will stop working. If you're having trouble using your fingers to navigate the touchscreen, you can buy a stylus. *(pen) that got that Rubber Tip*

Tap

Tapping is how you navigate your phone. If you want to launch an app, type on your keyboard or select an item – all you have to do is lightly tap.

Double-Tap

Use the double-tap option to select or launch things. You can also use it to zoom in or out.

Touch and Hold

If you touch and hold an item for a second or two, a small popup with extra options will appear.

Drag

Touch and hold an item and then drag it to a new place. You can use this option to drag apps and widgets to new locations.

Touch Sensitivity

If you want to protect your screen, it's important to get a screen protector. A screen protector will ensure that your screen does not crack if your phone falls or if something hits it. When you get a screen protector, go to the settings and change the touch sensitivity to make sure that the screen responsiveness to your touch is adjusted correctly.

[handwritten margin note: Store clerk did ~~^~~ Re Calibrate]

You can do this by:

1. Swipe up from the home screen and click on Apps.

2. Now go to Settings – Advanced features.

3. Find the Touch Sensitivity option.

Navigation Bar

There is a navigation bar at the bottom of the screen that will help you to control all the actions that you perform on your phone. By using this navigation bar, you can exit apps and return to your home screen.

The navigation keys are hidden when your phone is in full-screen mode. Even when the full-screen mode is on, you can still use the home button by clicking the place where the home icon generally appears when the navigation bar is on display.

If you want to enter or exit full-screen mode, all you have to do is:

1. Double tap on the Hide/Show icon that appears on the left of the navigation keys.

2. If you are already in full-screen mode and want to see the navigation keys, then swipe up

from the bottom of the screen where the navigation bar generally appears.

Tip: Navigation Bar Settings

You can change the navigation bar settings to make it more aesthetic and interactive. Multiple options are available in the settings, such as changing the background color and button layout.

1. Go to the Home screen, swipe up and click on Apps.

2. Tap Settings – Display – Navigation Bar. After this, you'll be able to access the following settings:

Show and Hide Button

You can add or remove the show/hide button that appears on the left of the navigation keys.

Background Color

You can change the background color of the whole navigation bar to make it more fun and to make your phone feel alive.

Button Layout

You can decide the order in which the navigation keys appear.

Home Screen

Your home screen is essential, and you have to configure it in a way that suits your needs. Your home screen will help you to access all the functions of your phone. Hence, you need to focus on how it looks, as well as how it works.

At the top of your home screen display, there is a Status Bar. This Status Bar tells you about the different notifications you have, calls or texts you have missed, the time, Wi-Fi Connections, and everything else. At the bottom, there's the navigation bar, and above it, you can see App shortcuts.

The most important thing about your Home Screen is that it has multiple screens that will help you to store different app shortcuts and widgets. You can drag the different shortcuts from one screen to another, delete them, edit them, club them together, and even create multiple screens.

If you want to align app shortcuts or even remove a home screen:

- Go to a home screen and pinch the screen. Once you have done that, you'll see different options to align apps to the top, bottom, or to even remove the screen itself.

- If you want to set a specific screen as your main Home Screen, click on the Main option. This screen will now be displayed when you click on the Home button.

- If you want to add a new screen, keep on swiping until you reach the last screen and click on Add.

App Shortcut ✓ Done

App Shortcuts allow you to access apps easily. By making a shortcut and keeping it on your Home Screen, you'll be able to access that app quickly from your Home Screen.

You can create an App shortcut by swiping up from the Home screen, clicking on Apps, and then scrolling to the app you want to make a shortcut of. After this, touch and hold the app and you'll see the Add to Home option.

If you want to remove an App shortcut, tap and hold the shortcut and a Remove option will appear. Removing the app shortcut won't delete the app. You can also switch the location of a shortcut by dragging it from one place to another or from one screen to another.

Wallpapers ✓ Done — Bullseye

You can change the background of your home and lock screen by setting up different wallpapers from the available ones on the phone.

- Go to a Home screen and pinch the screen.

- Click on Wallpapers to set a new one.

- Click on View All to see all the options.

- Scroll through the images and select the one you want.

If you select an image, you'll see an option asking which screen you want to apply it to. If you want a wallpaper to apply to all screens, pick from the Infinity Wallpapers. You can select Motion effect if you want to add movement to your screen.

Themes

Wallpapers are merely background images, while themes are more interactive. If you want to style your phone a certain way by making the status bar, icons, and other widgets all follow a single theme, and you can do that by using the Themes option.

1. Go to a Home Screen and then pinch the screen.

2. Click on the Themes option to set up a theme on your phone. When you tap on a theme, it will automatically be shown as a preview and then downloaded to My Themes.

3. You can see all the downloaded themes by clicking on View all.

4. Select the theme that you want and then click on Apply to apply the theme.

Just keep in mind that themes take a lot of battery life, since they are extremely interactive. If you are someone who uses their phone a lot and want to preserve your battery for as long as possible, it's better not to apply a theme.

Icons

All apps have built-in app icons that are shown as App shortcuts. By using the Icons menu, you can customize the icon of an app instead of using the built-in one. This can be particularly helpful if you're trying to hide an application from other people.

1. Go to a Home Screen and then pinch the screen.

2. Click on Wallpapers and then Icons. When you tap on an icon set, it will automatically be shown

as a preview and then downloaded to My Icons. *Test this*

3. You can see all the downloaded icons by clicking on View all.

4. Select the icon that you want and then click on Apply to apply the icon set.

Icon Frames: *Test this*

You can add side frames to icons if you want them to pop out on the home and app list screen. To do this:

- Swipe up from a Home Screen to access Apps.

- Go to Settings, Display, Icon Frames, and select from one of the following options:

- ➢ Icons only: Show icons only.

- ➢ Icons with frames: Show icons with shaded frame

- Click on Done and then follow the prompts.

Widgets *? what types to create?*

Widgets are small applications that you can run on your home screen. Widgets can be anything from a Google search bar to a clock. You can add as many widgets as you want to your home screen by following these steps:

1. Go to a Home screen and pinch it.

2. Click on Widgets.

3. Press and hold the widget that you want and then drag it to the home screen.

You can remove a widget by just pressing and holding it until the remove option appears on the Home screen.

Home Screen Settings

Like NOTE 5

Where is DRAWER to move Apps?

If you want to change your home screen settings, go to a home screen and pinch the screen. There are many different types of Home Screen Settings that you can customize and configure depending on your needs.

You can change the Home screen layout so that only one home screen appears with all the apps in it or the home screen and the app screen can be different. You can pick the one you want from different home screen grids; grids will allow you to change the way icons are organized on your home screen. You can similarly change the grid on the Apps screen to arrange the different App icons how you want. You can also add an Apps button to the home screen that will give you easy access to the Apps screen.

TURN ON OR OFF   CONTROL ? HOW SELECT

When apps <u>receive notifications</u>, a small <u>badge is added to the icon</u>. In the Home screen settings, you can also change the App icon badges to whatever you want. If you want app shortcuts to appear on the Home screen when you download them automatically, you can also find this setting in this menu.

The Home screen setting has many other options as well, such as portrait mode, which prevents your phone from turning to landscape mode, and hide apps option. This setting menu is a great feature for those who like to customize their phone screens for easier accessibility.

Easy Mode

If you are someone who hates excessively styled home screens and want a simpler visual experience for your phone, you can use the Easy Mode. Easy Mode allows you to switch from the default screen visuals to a simpler version of it where the texts and icons are larger. To access Easy Mode:

1. From the Home screen, swipe upward to access Apps.

2. Press Settings – Display – Easy Mode.

3. Press on Easy Mode, and it will become enabled.

You can follow the same steps to disable Easy Mode and go back to the Standard Mode.

Fuck This ✓ *For idiots — Why Buy the phone $$$ Get A RAZOR*

Status Bar Tricks

The status bar appears at the top of the screen, and there are a few tricks that you can use to make it more efficient. To perform the tricks mentioned below, swipe up from the home screen to access

Apps and then go to Settings – Display – Status Bar.

1. Notifications tend to get clustered, and it becomes hard to sort through that mess. You can change the settings of the status bar so that it only shows you the three most recent notifications at a time. Click on Show Recent Notifications Only to do this.

2. If you want to look at what percentage of battery charge is left, click on the Show Battery Percentage option and you'll be able to see the percentage next to the battery icon on the status bar.

Notification Panel

The notification panel is a status bar that you can access by swiping down from the top of your phone display. It's an extremely useful panel because you can use it to do a lot of things – access settings, access some quick settings for Wi-Fi, sound, Bluetooth and more, and see notifications from apps through notification cards.

The <u>notification panel</u> is basically where you run your whole device. If you want to access it, swipe down from the top of your device. You can deal with notifications in multiple ways from the notification panel:

1. If you want to view a notification, tap it, and you'll be redirected to

the app the notification card was from.

2. You can clear a notification by dragging it to the left or right.

3. You can clear all the notifications at the same time by tapping Clear.

You can close the notification panel by dragging it back up or by tapping 'Back' on the navigation panel at the bottom. If you don't want to access the notification panel by swiping down from the top, you can also open or close it by swiping up or down on the fingerprint scanner. To enable this:

1. Go to a Home screen and swipe upward to access Apps.

2. Go to Settings – Advanced Features – Finger sensor gestures

3. Click on 'on/off' to enable the feature.

Edge Screen — *Explore this practice*

Look on YouTube for this

Edge Screen allows you to customize information from apps, calendar, weather, etc. so that it is easier to access. This way all of your important information can be quickly accessed on the Edge screen.

Edge displays information from your **five most recent** most used apps so you can access them at a moment's notice. You can also customize Edge to make sure that any feature of your phone that you tend to use a lot is easily available to you on the Edge screen. So, if you like taking selfies a lot, you can add shortcuts for your favorite apps that you use to take selfies.

Edge panel works when your device is in standby. If you swipe from the edge of the screen when the device is in standby, you'll be able to see all the functions or shortcuts you have added to your Edge screen. You can even add a <u>list of contacts</u> to the Edge panel – these contacts will light up in different colors when they call.

You can also access Edge panel handles by dragging it from any screen to the center, and then you can swipe left or right to view different panels.

If you want to configure Edge panels:

1. Go to any screen, take the Edge panel handle and then drag it to

the center of the screen. After this, tap the Edge panel settings.

2. All you have to do is tap on/off to enable or disable the feature. You will see the following options:

> ➤ Checkbox: Allows you to enable or disable each panel.

> ➤ Edit (if available): Use tapping and configure each panel.

> ➤ Download: Use the Galaxy Apps to search and download more Edge panels.

> ➤ More Options – Reorder: Allows you to change the order of the Edge panels.

You can reorder by dragging them left or right.

> More Options – Uninstall: Delete an Edge panel from your phone.

3. Press back, and the changes will be saved automatically.

Apps Edge *Look on YouTube for Demo*

Apps Edge allows you to add app shortcuts to your Edge panel for faster access to apps and notifications. The <u>maximum</u> number of apps that you can add to the Apps Edge panel is <u>ten in two columns</u>. To do this:

1. Go to any screen, hold the Edge panel handle and drag it to the

center of the screen. Keep swiping until the Apps Edge panel appears on the screen.

2. Just tap on an app or app pair if you want to open it.

Configuring Apps Edge can be done by:

> ➤ Go to any screen, hold the Edge panel handle and drag it to the center of the screen. Keep swiping until the Apps Edge panel appears on the screen.

> ➤ Find Add app option and tap it to add other apps to the Apps Edge.

> ➤ Add an app to the Apps Edge by finding it on the left side of the screen. Tap the app and then add it to any of the available spaces in the right column.

- If you want to create a shortcut so that two apps can open in multi-window, find and tap on Create App Pair. You can create a folder shortcut by taking an app that is on the left side of the screen and dragging it on top of an app in the right column.

- You can change the order of the apps on the Edge panel by holding and dragging the apps to your preferred location.

Remove an app by simply tapping on Remove.

3. Tapping back will automatically save any changes.

YouTube

People Edge   *This is New*
Do This!   *I Need to setup*

People Edge allows you to set contacts for your Edge screen so that you can easily and quickly talk to any of your contacts from the My People list.

This Book Needs Pictures!! WTF!

1. Go to any screen and then drag the Edge panel handle to the middle of the screen. Continue swiping until the People Edge panel comes.

2. Find the contact and tap it. You'll get two options after this – Call or Message, tap whichever you want.

If you want to configure People Edge:

1. Go to any screen and then drag the Edge panel handle to the

middle of the screen. Continue swiping until the People Edge panel comes.

2. Find and tap the Add Contact option.

   ➢ Add a contact by tapping on Select Contacts.

   ➢ Remove a contact by simply tapping Delete.

   ➢ Change the order of the contacts by dragging the contacts to your preferred locations.

3. Tapping Back on the navigation panel will automatically save the changes.

What is Edge Panel Handle

Smart Select

S9 and S9+ both come with the amazing Smart Select feature, which allows you to capture a specific area on the screen of your device in the form of an animation or image. You can do anything that you want with this animation or image – share it with your friends or pin it to the screen. To access the Smart Select feature:

1. Go to any screen. Find the Edge panel handle, hold and then drag it to the middle of the screen. Keep swiping the panels until the Smart Select panel comes.

2. There are multiple Smart Select tools that you can use. Pick from one of the following:

> Rectangle: Takes a rectangular view of the screen and captures it.

> Oval: Takes an oval view of the screen and captures it.

> Animation: Creates an animated GIF of whatever is on your screen.

> Pinto screen: Captures a certain area and then pins it to the device's screen.

Edge Lighting

Edge Lighting is a setting that you can turn on which makes your Edge screen light up if you receive calls or notifications. This function only works when your phone is turned over. To enable Edge Lighting:

1. Go to a Home screen and then swipe upward to access Apps.

2. Now go to Settings – Display – Edge Screen – Edge Lighting.

3. Press the on/off button to enable or disable this feature.

4. You'll see multiple options that allow you to set different Edge Lighting settings. Pick from the following:

When screen is on: Pick this option if you want Edge Lighting to work only when the screen is on.

When screen is off: Pick this option if you want Edge Lighting to work only when the screen is off.

Always: Pick this option if you want Edge Lighting to function at all times.

Edge Lighting Advanced Settings

If you want Edge Lighting to be even more fun, you can go to the Edge Lighting Advanced Settings and change the color, transparency, and even width of the Edge Lighting feature.

1. Go to a Home screen and then swipe upward to access Apps.

2. Go to Settings – Display – Edge Screen – Edge Lighting.

3. Tap the Edge Lighting style option. In this setting, you can customize - *Effect: Pick an edge effect.*

   ➢ Color: Pick one of the preset colors or customize yourself. You can also enable and pick app colors.

   ➢ Transparency: You can adjust the transparency of the Edge Lighting by dragging the transparency slider.

   ➢ Width: Just like the transparency slider, drag the width slider to change the width of the Edge Lighting.

4. After you are done, just press Apply.

Manage Edge Lighting Notifications

In this setting, you can pick and choose the notifications that will light up on the Edge screen. To customize the notifications on the Edge screen:

1. Go to a Home screen, swipe up and then access Apps.

2. Go to Settings – Display – Edge screen – Edge Lighting.

3. Find and press Manage Notifications to determine which apps' notifications will light up the Edge Lighting feature.

Quick Reply

If you have an incoming call, you can quickly reject it and send a text message to the caller by pressing your finger down for two seconds on the heart rate sensor. The call will automatically be rejected, and a preset message of your choice will go to the caller. To set Quick Reply:

1. Go to a Home screen and then swipe upward to access Apps.

2. Go to Settings – Display – Edge screen.

3. Go to Edge Lighting – More options – Quick reply.

You can edit the reply text by tapping on the already existing message.

Chapter Four: Apps

There are many apps on the Samsung Galaxy S9 and S9 plus. Most of these apps have been pre-loaded for your accessibility and the rest you can download from the Google Play store.

In this section, we will discuss the most important apps on your device, how you can access and navigate them, and the features of these apps that truly make them special.

Before that, we need to learn how to access apps in general. If you want to access apps, all you have to do is go to any Home screen and swipe up or down to show apps. You can also create app shortcuts as mentioned above and can access these app shortcuts from where you have placed them on the Home screen. You can also create an Apps button on the home screen if you don't want to swipe up or down every time to access apps. To do this:

1. Go to a Home screen and swipe upward to access Apps.
2. Press More Options – Home Screen Settings – Apps button.
3. Now press Show Apps button – Apply.

Any changes that you want to do to an app, such as deleting it or disabling it, you can do by going to the Apps list from your home screen and then pressing and holding the concerned app.

Camera and Video

The S9 and the S9 plus have great cameras that work with dual-aperture, which means that the camera lens adjusts itself depending on the amount of light that is entering the lens. This makes sure that every picture or video that you take looks perfect and is of high quality.

To take photos and to capture videos, open the camera app by going to the Apps list from your home screen. You will now see a display screen that shows you a preview of the picture you are going to take. When you are taking a picture, there are three options available:

i. Tap the screen once to focus the shot. Once you have done that, a brightness slider will appear on the screen. Reduce or increase the brightness to adjust the picture.

ii. If you want to change the shooting mode to something else, swipe the screen left or right.

iii. You can change between the front and the rear cameras quickly by swiping the screen up or down.

Just tap the small circular button to take a picture or press one of the volume buttons on the side. You can similarly record a video by tapping the record option that appears on the preview screen. While in video mode, tap on pause or stop to navigate through the video making process. You can even take a photo while the video is being recorded – click on the capture button.

Shooting Mode

Both the front and rear cameras offer lots of shooting modes that allow you to take different kinds of pictures. You can enable different shooting modes by swiping to the left or right from the main preview screen. The different shooting modes available are:

Rear Camera

Food: With this, you can take pictures that will make the colors of food pop out.

Panorama: Move your phone horizontally or vertically to capture a larger area while creating a linear image.

Pro: You should use this setting if you have some professional experience taking photos. In this mode, you can adjust the ISO Sensitivity, white balance, colors, and exposure.

Selective Focus: You can change what kind of focus each part of the image receives after you have captured it with this mode. You can change focus to the subjects far away or near to the camera. This feature is not available on the S9+.

Auto: This feature allows the camera to adjust the focus and other aspects of the photo itself.

Super slow-mo: You can record videos and then play them at a super slow frame rate for a great video effect.

AR Emoji: You can create an emoji for yourself or even things around you. AR Emojis can also be animated using the AR Emoji configuration.

Hyperlapse: With this, you can create a video that quickly skips through time to create a time-lapse video. You can't adjust the time/rate given to each frame as it is automatically determined by the movement of your phone and what is being recorded.

Slow Motion: This feature allows you to record a video at a high speed and the video is then played in slow motion.

Sports: If you want to take pictures of fast-moving objects, use this mode. It is great for taking pictures of people playing sports.

Front camera

Selfie Focus: If you want your face to pop out in a selfie, then use this mode to blur the background and push all the focus to your face.

Selfie: This feature allows you to edit your selfies. You can add different effects like airbrush, etc.

AR Emoji: Similar to the rear camera, this feature allows you to create an Emoji of yourself by taking a selfie.

Wide Selfie: If you have a huge group of people who you want to take a selfie with, then use this mode to fit them all in the picture.

Top 5 Camera Features

Galaxy S9 and S9+ have amazing cameras featuring a pair of 12-megapixel lenses with wide-angle shooting and optical image stabilization (OIS) system. Thanks to all of this, the cameras have some amazing features, such as:

1. Live Focus: This mode will allow you to take different artistic photos by changing the depth of field at any point of time in the photo taking process. This feature is sadly not available for the S9.

2. Background Blur: The background blur option allows you to blur the background and add shapes like hearts, twinkles, and stars to it.

3. Camera Aperture: Although the dual aperture system of the S9 and S9+ are both controlled automatically by the camera, you can adjust the camera aperture yourself too. This will allow you to change the light entering the camera so that the image can adapt better based on the environment. To do this go to the pro mode, swipe left twice and press the aperture icon. If you are in landscape mode, you will see the camera aperture that the

camera currently is on at the bottom of the screen.

4. Create GIFs: The S9 and S9+ camera allow you to create slow motion videos that can be used as GIFs. Just switch on the super slow-motion feature, which increases the camera speed to 960 frames per second – this means that the video you are recording is four times slower than a normal video rate. You can do anything that you want with the slow video you have recorded. You can make a GIF out of it, increase the speed of certain parts, add it as wallpaper on the home and lock screen, or add music to it.

5. Viewfinder Grid Lines: If you want to improve your phone photography, the best thing that you can do is enable viewfinder grid lines. These grid lines will allow you to get perfect shots by aligning the horizon and making the images proportionate. You can enable grid line by going to the camera settings. Just pick from the multiple options given in the list; the "3 x 3" option works best, but Samsung has introduced a new option called Square, which is also great.

Galaxy Apps

Remember to download all the Samsung Apps that are essential for your device. These are known as Galaxy Apps, and they will help your phone to function better. You can download these apps by swiping up from your home screen to access Apps and then going to more options and tapping Galaxy essentials.

You can access Galaxy Apps by swiping up from your home screen and going to the Apps section. Click on the Samsung Folder and find the Galaxy Apps option.

Gallery

You can view photos and videos that you have taken or the ones you have received in the Gallery. The Gallery is important because it allows you to edit your pictures so that they look great.

Just tap on a picture to view it, and you will see the editing options. You can transform your photographs by rotating them or flipping them, add color effects, add animated stickers, and even draw on pictures directly on them.

Samsung Health

Samsung Health is a great app that you can use to track your health-related activities. You can track your meals, the number of calories and other nutrients you are taking, manage exercise routines, and even monitor your sleep pattern.

Samsung Health allows you to track your heartbeat as well and can be the replacement for expensive devices that track these things for you. Although, it is possible that the readings might not be as reliable so please contact your physician before you decide to use the app as your main source for tracking your health-related activities.

You can access the Samsung Health app by swiping up from a Home screen and going to the Samsung Folder. In the Folder, find the Samsung Health app and tap on it. You will be asked to read the terms and conditions, after which you will have to enter your health details to start the app.

Chapter Five:

Special Features

In the first couple of chapters, we looked at the different features of the S9 and S9+ that make them the best phones on the market. In this chapter, I will tell you the most important features of the S9 and S9+ that will make sure that your Android experience is as good as it can be.

Bixby

Bixby is the virtual assistant that Samsung has created for your phone. It's just like Amazon's Alexa or Apple's Siri and it works via voice command like any other virtual assistant. Bixby learns from whatever command you give it and evolves so that it can fulfill all of your needs. Bixby had a slow start when it was first launched with the Samsung Galaxy S8, but with the S9 and S9+, Bixby has genuinely evolved and can now do much more with the information you provide it.

All you need to do is tell Bixby what to do, and it will help you with it – from calling people, Googling information or opening Apps, it can do it all. Bixby's vocabulary is still in development and is not as good as Apple's Siri, but it does the job.

The Bixby button is on the left of your device, and you can press it to enable it. Once you press the button, you can use

Bixby by giving it a voice command or telling it what you're interested in.

Bixby has improved a lot with the S9 and S9+, and you should start by using Bixby in the camera app. Go to the camera app and tap on the Bixby vision icon – it looks like an eye and is not that difficult to find. Bixby will now open inside the camera app, and you'll be able to see the eight different Bixby vision modes that range from food to makeup. You will be able to see these vision modes at the bottom of the screen, just below the camera's preview screen.

Let's first start with the basics of Bixby and then we'll move on to discuss the different tips and tricks related to Bixby.

Setting Up Bixby

Bixby works via voice command, and you can give a voice command to Bixby by simply pushing the Bixby button on your device. But, before you do all of this, you have to register your new S9/S9+ device with the larger Bixby framework. This will help Bixby to keep an account with your information, so it can learn from it and evolve to serve your needs quickly and more efficiently. Bixby can even evolve itself to a point where it will start anticipating what you need and giving you prompts to fulfill your desires.

The first thing you will have to do is tell Bixby your preferred language – there are only three options available: English,

Mandarin, and Korean. Samsung is behind even in this, since Siri gives you the option to select from more than 20 languages. After you have set the language, you will have to go through a little voice training to make sure that Bixby understands your voice. Bixby is great at replying to the most obvious and simple commands like 'Open this app' or 'Call mother.' Bixby has a list of preset commands that it functions with, and if you want to use it effectively, you will have to follow them. You can find the list of commands on Samsung's website.

Bixby has a lot of functionality, and with it, you can ensure that you're not endlessly tapping at the screen all the time. Bixby allows you to do the most basic things like setting up events, alarms, or giving you reminders. It can also be used to launch cameras, send emails, and open other apps like Google or Maps. Bixby works best with in-built Samsung apps that you can direct using Bixby. Bixby also works effectively with a few third-party apps such as Twitter, Instagram, and Facebook – you can run these apps without ever touching your screen with the help of Bixby.

Bixby is not that great as a search tool. Bixby will be able to give you obvious answers, but if you ask it something complex, it gets confused and is not able to answer. Google's assistant is still better at this job, and I would recommend that you use it.

Bixby Vision

Bixby is different from all the other artificial intelligence assistants because of Bixby Vision. Bixby became famous when it was launched with the S8 because it could integrate well with the camera app, and by just pointing the camera towards a book, Bixby could bring up the price on a site like Amazon.

Bixby Vision has taken a massive leap with the S9, and it has become even better. The best feature is that Bixby can translate street signs and even menus into 53 different languages.

It barely takes a few seconds for the translation to happen – open the camera app, click on the Bixby eye vision icon, pick the translation option, and point the camera towards what you need translated. The translation is not perfect, and at times Bixby tends to give the wrong answers so be careful until Samsung rolls out another update.

Here are all the other things you can do with Bixby Vision:

> Scan QR Codes: Bixby works extremely well with QR codes. The QR code will appear as you keep swiping right through the Bixby vision options. Bixby will scan the QR code and send you to the requisite website.

> Wine rating: This is a fun feature as it allows you to scan the name of the wine that you're drinking. Just pick the option that looks like a wine glass and point it

towards the label of your wine bottle. Soon, you'll get the rating and vintage of the wine that you're drinking. You can even get more details such as food that pairs well with your wine.

- Makeup: Bixby can take a live selfie and display various makeup products on it. This feature is extremely useful, and it will allow you to select from different makeup products by testing them on your face.

- Converting Images to text: If you have an image of a letter or even handwritten notes that you want converted to text, Bixby is the way to go. Bixby's converter is pretty efficient and can convert images into text with accuracy.

Top 5 Hidden Bixby Features

Bixby has many hidden features that are not that well publicized. In this section, we will look at the top 5 features of Bixby and how to use them.

1. Samsung Rewards – As you continue to use Bixby you will earn XP just like in a video game. You can then use the reward points that you earn to redeem them for gifts. You can do this by clicking on the Bixby key and a prompt will appear that says, "*Join the Bixby promotion for Samsung Rewards.*" All you have to do now is click on this link and register yourself. The XP can be used for Samsung Pay credits and can even be used to unlock different customization options for your Bixby.

2. Change listening sensitivity – It's no secret that Bixby does not work perfectly, and it does not always listen to commands. If Bixby does not hear you properly, you can fix it by changing the listening sensitivity of Bixby. You can do this by tapping on the Bixby button – Menu – Settings – Voice wake-up. Now, slide the sensitivity bar to medium or high.

3. Jokes – Bixby, like any other virtual assistant, is great at making people laugh. If you want Bixby to tell a joke, use the following commands:

> "Rap for me."
>
> "Tell me a joke."
>
> "Beatbox for me."
>
> "I love you."
>
> "Ok, Google."

4. Make Bixby your secretary – Bixby can take notes for you just like a secretary. When your note-taking app is open say, "Hi, Bixby, dictate," and say whatever you want Bixby to take note of.

5. Search Emails – Bixby allows you to search your emails for mail from a specific person. You can even search your emails for a specific mail that you have been looking for. Just tell Bixby to search emails from xyz person, and it will only take a few seconds.

Securing Your Phone

The one thing that makes the S9/S9+ stand out is the security options that the devices offer. Securing your phone is essential in this day and age, especially because everything about us can be deduced by studying our phones. Samsung has truly gone out of its way to ensure that your security is guaranteed with the S9/S9+.

In this section, we will look at the multiple security options that the S9/S9+ have to offer, how you can enable them, and the advantages and disadvantages of each one of them.

Face Recognition

Face recognition allows you to use your face as a form of security check when you are opening your phone. Face recognition can also be used to lock different apps; if you are worried about any apps that people might try to use, just set up face recognition. Face recognition is safer than a password since it can't be guessed. Face recognition on the S9/S9+ works pretty well, and the front camera scans your face to unlock the screen.

The first thing that you will have to do is register your face to access this feature. Before you set face recognition, you will have to set a PIN, pattern or password as an alternative way to access your phone. If you want to set face recognition, you can register your face by following these steps:

1. Go to a Home screen, and then swipe upward to access Apps.

2. Now go to Settings – Lock Screen and Security – Face Recognition.

3. You will now see prompts on your screen. Just follow these prompts and register your face.

Face recognition has even more detailed features, and if you want to you can customize face recognition as well. Just follow these steps:

➢ Go to a Home screen, and then swipe upward to access Apps.

➢ Go to Settings – Lock Screen and Security – Face Recognition.

Now, you will see the following options:

Remove Face Data: Use this option if you want to delete the current face that is set as the lock.

Samsung Pass: This option allows you to access your online accounts via face recognition.

Face Unlock: This option allows you to turn face recognition on or off.

Face Unlock when Screen Turns On: This option allows you to open your phone with face recognition whenever the screen is enabled.

Faster Recognition: This option makes face recognition faster. You can turn this option off if you're worried about someone using a video or image of you to unlock your phone.

Iris Scanner

Iris Scanner works the same way as face recognition; the only difference is that instead of scanning your face, your irises get scanned. You can use the iris scanner to unlock your phone or even to unlock apps. It's an alternative to using passwords, patterns or PINs.

The first thing that you will have to do is register your irises with the device. Follow these steps:

1. Go to a Home screen, and then swipe upward to access Apps.

2. Now go to Settings – Lock Screen and Security – Iris Scanner.

3. You will now see prompts on your screen. Just follow these prompts and register your face.

There is also an option to register just one of your eyes. To do this, go the setup screen and tap the link that prompts you to register just one iris.

Iris Verification allows you to access other accounts that are linked with your device with the help of iris verification.

- ➢ Go to a Home screen, and then swipe upward to access Apps.

- ➢ Go to Settings – Lock Screen and Security – Iris Scanner. Now, you will see the following options:

Remove Iris Data: Use this option if you want to delete the current face that is set as the lock.

Preview Screen Mask: You can pick a mask that will be displayed on the screen when you are using the iris scanner.

Samsung Pass: This option allows you to access your online accounts via iris recognition.

Samsung Pay: This option allows you to make online payments via iris recognition. It's a secure option and is also very quick.

Iris Unlock: Turn this on if you want to unlock your phone using iris recognition.

Iris Unlock when Screen Turns On: This option allows you to open your phone with iris recognition whenever the screen is enabled.

Tips for Using Iris Recognition: Tap this, and you'll be able to see a tutorial that gives you tips on using iris recognition.

Fingerprint Scanner

Fingerprint scanning has become pretty standard for Android phones, and every phone has to have it now. You can use fingerprint scanning to not only open your phone but also to access the different accounts that you have linked with your device.

The first thing you need to do is register your fingerprint, but before that make sure that you have set a PIN, pattern or password first. There are three fingers that you can register; they can be your fingers or someone else's.

If you want to set fingerprint recognition, you can register your fingerprint by following these steps:

1. Go to a Home screen, and then swipe upward to access Apps.

2. Now go to Settings – Lock Screen and Security – Fingerprint Scanner.

3. You will now see prompts on your screen. Just follow these prompts and register your fingerprint.

After you are done registering your fingerprint, you can also access fingerprint management to make any changes that you want, such as adding, deleting, or renaming fingerprints.

> ➢ Go to a Home screen, and then swipe upward to access Apps.

> Now go to Settings – Lock Screen and Security – Fingerprint Scanner. Now, you will see the following options:

Renaming: You can give the fingerprint a new name by tapping on that fingerprint, entering a new name and just pressing Rename.

Adding: You can add up to three fingerprints. To add one, press Add fingerprint, and then some prompts will appear on the screen. Just follow these prompts, and you're done.

Deleting: The delete option for a fingerprint can be accessed by pressing and holding a fingerprint. You will see the Remove option – tap it.

You can also use your fingerprint as a form of verification to open other apps and to lock certain aspects of your phone.

> Go to a Home screen, and then swipe upward to access Apps.

> Go to Settings – Lock Screen and Security – Fingerprint Scanner. Now, you will see the following options:

Samsung Pass: This option allows you to access your online accounts via fingerprint recognition.

Samsung Pay: This option allows you to make online payments via fingerprint recognition. It's a secure option and is also very quick.

Fingerprint Unlock: Turn this on if you want to unlock your phone using iris recognition.

Intelligent Scan

Intelligent Scan provides you with complete security by combining both iris and face recognition. It truly makes it impossible for people to access your phone unless you want them to. It not only improves security, but it also improves accuracy.

You can register your face and iris by following the steps given above. You can customize how Intelligent Scan works by following these steps:

1. Go to a Home screen, and then swipe upward to access Apps.

2. Now go to Settings – Lock Screen and Security – Intelligent Scan. You will now see the following options:

Samsung Pass: This option allows you to access your online accounts via intelligent scan verification.

Intelligent Scan Unlock: This option allows you to turn face and iris recognition on or off.

Screen-on Intelligent Scan: This option allows you to open your phone with intelligent scan whenever the screen is enabled.

Remove Face and Iris Data: Use this option if you want the current face and iris information to be deleted.

Multi Window

Multi Window allows you to split your display screen into two so that you can use two apps at the same time. This feature allows you to multitask two apps at the same time. Multi Window is a great feature, but it's one that has been around for some time on Android phones. Samsung has changed its Multi Window feature so that you can easily switch between apps, and you can even change the size of the windows according to your needs.

1. On any screen, use the navigation panel and tap Recent Apps.

2. You can launch apps into split screen by tapping Multi Window in the title bar. Here are a few things that you should note:

   ➢ The app that was recently opened is displayed on the bottom screen. You will know which apps support Multi

Window and which don't by looking out for the Multi Window icon in the title bar.

➢ If there are no recent apps that can be used in Multi Window, or if you want an app that is not on the recent apps list, tap App List.

3. Tapping an app that is in the bottom window will add it to the Multi Window view.

Samsung Pay

Samsung Pay has grown since it started to pair up with all the major banks and even card companies like American Express and MasterCard. It also has a rewards program that is pretty great, and you can easily get rewards by using Bixby. You will have to set up the service first if you want to use it.

You can set it up by tapping the Samsung Pay icon and then you will be directed to registration. You will first have to log in to your Samsung account and, after that, you will need to add your Credit or Debit Card details, which you can easily do by using the S9/S9 plus' camera. The last thing you will have to do is add the verification code that your bank will send you.

Customizing your Screen

Both the S9 and the S9+ have great screens that give you a wonderful viewing experience. To do this, all you have to do is go the Settings menu and then tap the Display option.

From here, you will see multiple options that will allow you to set the way your screen looks. The best one is Adaptive Display – it will enable you to customize the color temperature of your screen.

There are other great options that you can try: AMOLED Cinema, AMOLED Photo, and Basic mode. Basic mode is the one you should use if you want a more natural feel to your phone and want the color saturation to be low. There is also a slider, which allows you to reduce or increase the color saturation.

AR Emoji

AR Emoji is a fun little feature that Samsung has developed with the S9 and S9 plus. AR Emoji is just a small little avatar of yourself that you can create by using the camera app. You can customize this AR Emoji to look like whatever you want, and you can then send it to people while you're texting them.

To access the AR Emoji feature, all you have to do is open the camera app and then open the front-facing camera. From the options at the top, select the AR Emoji option. You can then take your photo, which the phone will turn into an Emoji. You can also customize this Emoji to look like whatever you want.

Customize Messages App

Texting is an extremely interactive activity and the app that you use to text matters a lot. You can customize your messages app on the Samsung Galaxy S9/S9 plus to look like whatever you want. You can change the theme of the app or the font size to look however you want.

If you want the Messages App to have a different theme, change the theme of the whole phone. You can do this by following the instructions that were given in the earlier chapters. You can adjust the font settings by going to the Settings menu in the Messages App.

Conclusion

With that, we have come to the end of this book. Thank you once again for choosing this book, and I hope you were able to get all the information that you wanted about the Samsung S9 and s9+.

NOT REALLY

The primary purpose of this book is to be a reference guide for running the S9/S9 plus. So, if you ever feel like you are stuck on something, consult this guide again. *HA! Better to watch YouTube*

This guide can help you with the phone only to a certain extent; you will be able to learn everything about your phone only when you continue to use it regularly. Go through the different apps and settings to experiment with what your device has to offer.

It's imperative that you maintain the quality of your phone to make sure that it continues to function well for years to come. Your phone is protected against water and dust, but still, make sure that you do not take it near any dangerous substances. Remember to protect your phone by purchasing a screen guard. Also, don't tamper with your phone unless you know what you are doing.

Thank you for buying this book.

Made in the USA
San Bernardino, CA
21 July 2018